MIND BLOWING! THE BRAIN

SLEEP

by Joyce Markovics

Cherry Lake Press
Ann Arbor, Michigan

![Cherry Lake Press logo]

CHERRY LAKE PRESS

Published in the United States of America by Cherry Lake Publishing Group
Ann Arbor, Michigan
www.cherrylakepublishing.com

Reading Adviser: Beth Walker Gambro, MS Ed., Reading Consultant, Yorkville, IL
Content Adviser: Mark W. Green, MD, Neurologist
Book Designer: Ed Morgan

Photo Credits: © freepik.com, cover and title page; © freepik.com, TOC; © HBRH/Shutterstock, 4; © Olga Popova/Shutterstock, 5 top; © Chursina Viktoriia/Shutterstock, 5 bottom; © freepik.com, 6; © freepik.com, 7; © EA Photography/Shutterstock, 8; © Andrey_Popov/Shutterstock, 9; © freepik.com, 10; © freepik.com, 11; © freepik.com, 12; © SciePro/Shutterstock, 13; © freepik.com, 14; © freepik.com, 15; © freepik.com, 16; © freepik.com, 17; © freepik.com, 18; © freepik.com, 19; Luigi Novi/Wikimedia Commons, 19 inset; © freepik.com, 20; © freepik.com, 21; © freepik.com, 22.

Copyright © 2022 by Cherry Lake Publishing Group
All rights reserved. No part of this book may be reproduced or utilized in any form or by any means without written permission from the publisher.

Cherry Lake Press is an imprint of Cherry Lake Publishing Group.

Library of Congress Cataloging-in-Publication Data

Names: Markovics, Joyce L., author.
Title: Sleep / by Joyce Markovics.
Description: Ann Arbor, Michigan : Cherry Lake Publishing, [2022] | Series: Mind blowing! the brain | Includes bibliographical references and index. | Audience: Grades 4-6
Identifiers: LCCN 2021035293 (print) | LCCN 2021035294 (ebook) | ISBN 9781534199583 (hardcover) | ISBN 9781668900727 (paperback) | ISBN 9781668906484 (ebook) | ISBN 9781668902165 (pdf)
Subjects: LCSH: Sleep—Juvenile literature. | Sleep—Physiological aspects—Juvenile literature.
Classification: LCC QP425 .M344 2022 (print) | LCC QP425 (ebook) | DDC 612.8/21—dc23
LC record available at https://lccn.loc.gov/2021035293
LC ebook record available at https://lccn.loc.gov/2021035294

Printed in the United States of America
Corporate Graphics

CONTENTS

Sleepwalker 4
Your Brain and Sleep 8
Sweet Dreams 16
Sleep Issues 20
   Brain Games 22
   Glossary 23
   Find Out More 24
   Index 24
   About the Author 24

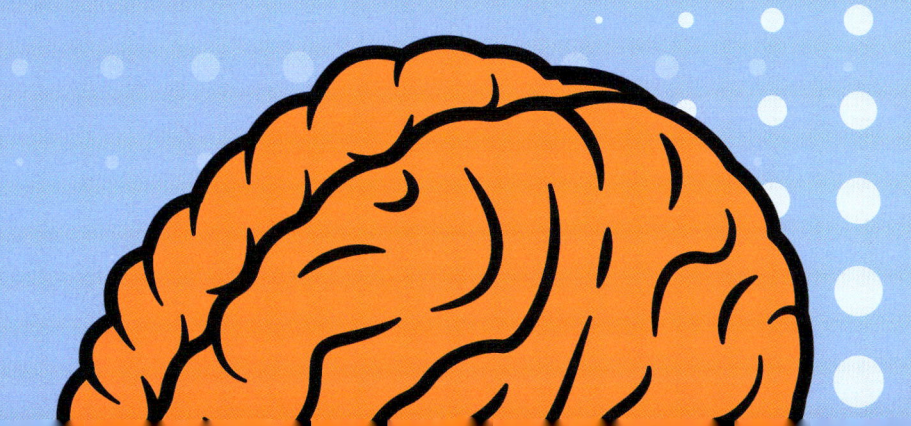

SLEEPWALKER

"I was first caught sleepwalking at eight years old," says Eleanor, who's from England. Eleanor would leave her bed in the middle of the night. In the morning, her mom would find her curled up under a heap of clothes in a closet. Eleanor's sleepwalking didn't stop there.

If you ever find someone sleepwalking, you shouldn't wake them. Try to gently guide them back to bed.

When Eleanor was older, she sleepwalked into her kitchen one night. A friend found her going through the freezer looking for chicken nuggets. "Don't just stand there, help me find them!" Eleanor barked—all while fast asleep. Even stranger, Eleanor is a **vegetarian**.

Eleanor doesn't remember when she sleepwalks. "Those were my most dreamless nights," she said. "When I woke up, I would be quite **disoriented**."

Eleanor often sleepwalked with her eyes open and while talking. It startled her friend. "He told me I looked like a **demonic** doll." Sleepwalking, or somnambulism (sohm-NAM-byuh-lih-zuhm), is when people get up and walk around while asleep. Some people will even drive when they're sleepwalking!

One in 50 adults suffers from sleepwalking.

"A sleepwalking person walks or makes other movements that appear to have a purpose," says sleep expert Nancy Foldvary-Schaefer. "We aren't sure of the exact cause." However, experts do know that one **organ** is responsible for sleepwalking and sleep—your brain!

A human brain looks like a pinkish-gray cauliflower and has a soft texture like pudding!

Long ago, people believed sleep was the result of gases rising from inside the stomach.

YOUR BRAIN AND SLEEP

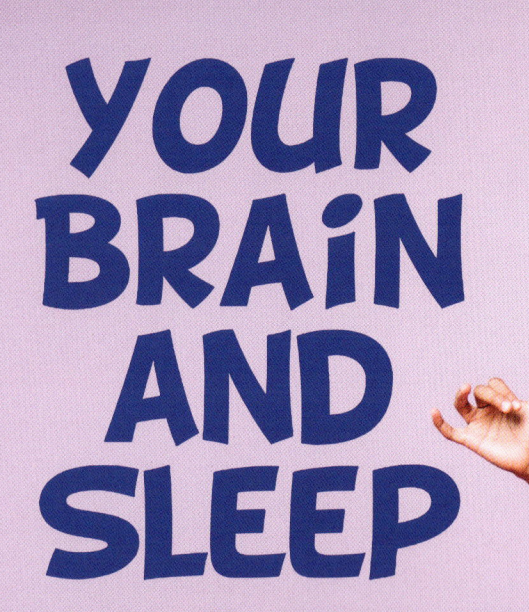

Your brain is the busiest organ in your body. It weighs only about 3 pounds (1.4 kilograms), yet it controls everything you do. From breathing and eating to jumping and reading, your brain directs every muscle and organ in your body. Without it, you wouldn't be able to move, think, remember, sleep, or dream!

Inside your brain are billions and billions of cells called neurons (NOO-ronz). **Synapses** link neurons together. Every second, neurons send **chemical** and **electrical** signals racing around your body. These signals allow you to leap over a puddle, laugh at a joke, or do anything at all! In order to function, your brain uses up lots of energy.

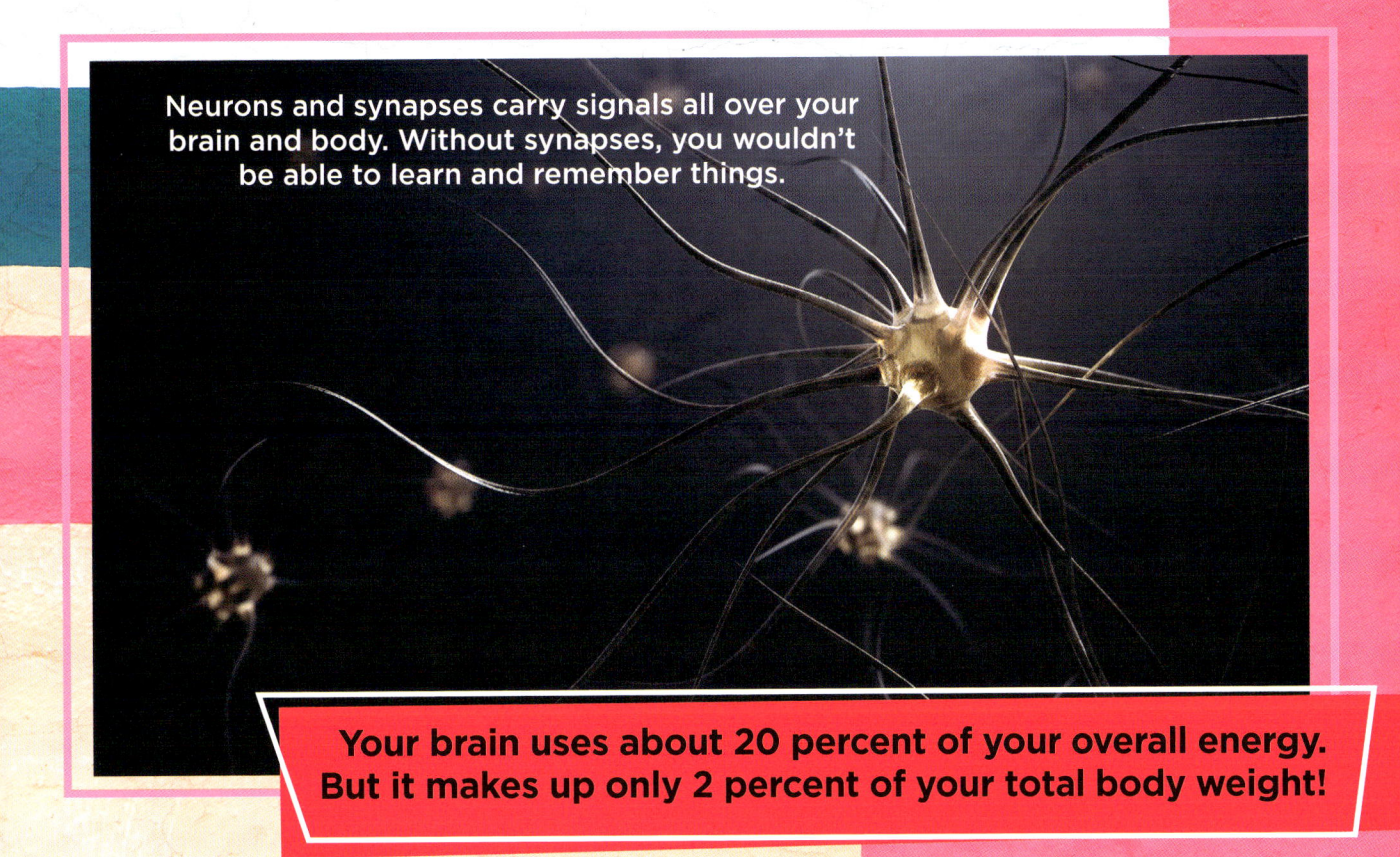

Neurons and synapses carry signals all over your brain and body. Without synapses, you wouldn't be able to learn and remember things.

Your brain uses about 20 percent of your overall energy. But it makes up only 2 percent of your total body weight!

Like a mini-vacation, sleep gives your brain and body time to rest. In fact, every animal on Earth needs sleep to survive. Think about how much a cat or dog naps. Animals sleep a lot—for the same reason you do! Your body, especially your brain, needs a certain amount of sleep every day.

Cats sleep for 12 to 16 hours each day. Lions sleep even more—at around 20 hours per day!

The amount of sleep you need ranges from 7 to 11 hours, depending on your age. Experts aren't sure what exactly the brain does while you're asleep. However, they know it's still working. Scientists think the brain sorts and stores information, saving what's important and tossing what isn't. The brain also **replenishes** chemicals and cleans out waste products during sleep.

> Kids need more sleep than adults—about 9.5 to 11 hours each night. Newborn babies can sleep up to 17 hours per day.

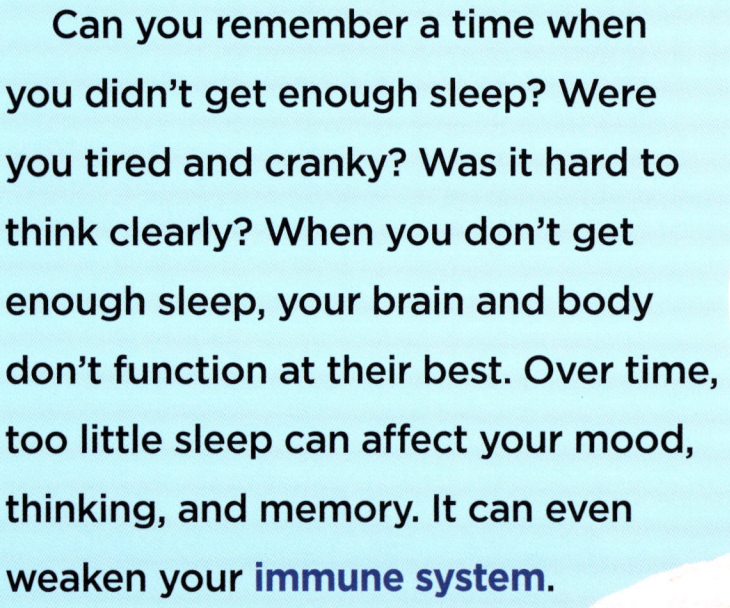

Can you remember a time when you didn't get enough sleep? Were you tired and cranky? Was it hard to think clearly? When you don't get enough sleep, your brain and body don't function at their best. Over time, too little sleep can affect your mood, thinking, and memory. It can even weaken your **immune system**.

A person can survive for 3 weeks without food and water but die from lack of sleep after about 10 days.

Sleep has different stages. When you're falling asleep, it might not feel like much is happening. However, your brain is doing a lot. Brain parts called the brain stem, **hypothalamus**, and cortex are telling your body how to sleep. The cortex is the gray, wrinkly **tissue** covering your brain.

cortex

hypothalamus

brain stem

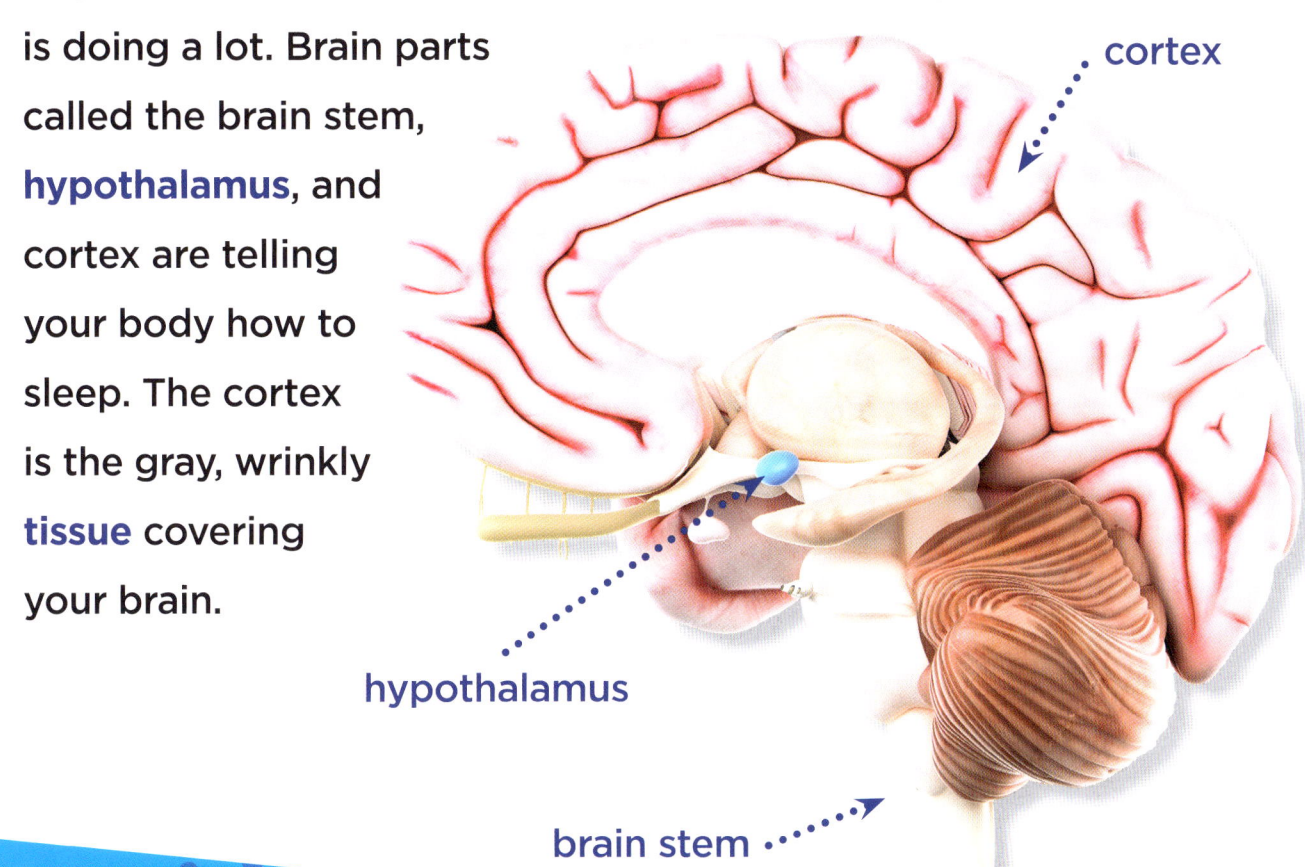

The brain stem controls many basic body functions, such as breathing and swallowing, without your having to think about them!

In the first stage of sleep, you can be awakened pretty easily. In the second and third stages, your brain signals your muscles to relax. Your heart rate and breathing slow down. At this deeper stage of sleep, it can become hard to wake a person. And the brain becomes a little less active.

5 Stages of Sleep

1. **DROWSY** 5–10 minutes
2. **LIGHT SLEEP** 20 minutes
3. **MODERATE SLEEP** 20–40 minutes
4. **DEEP SLEEP** 30 minutes
5. **RAPID EYE MOVEMENT**

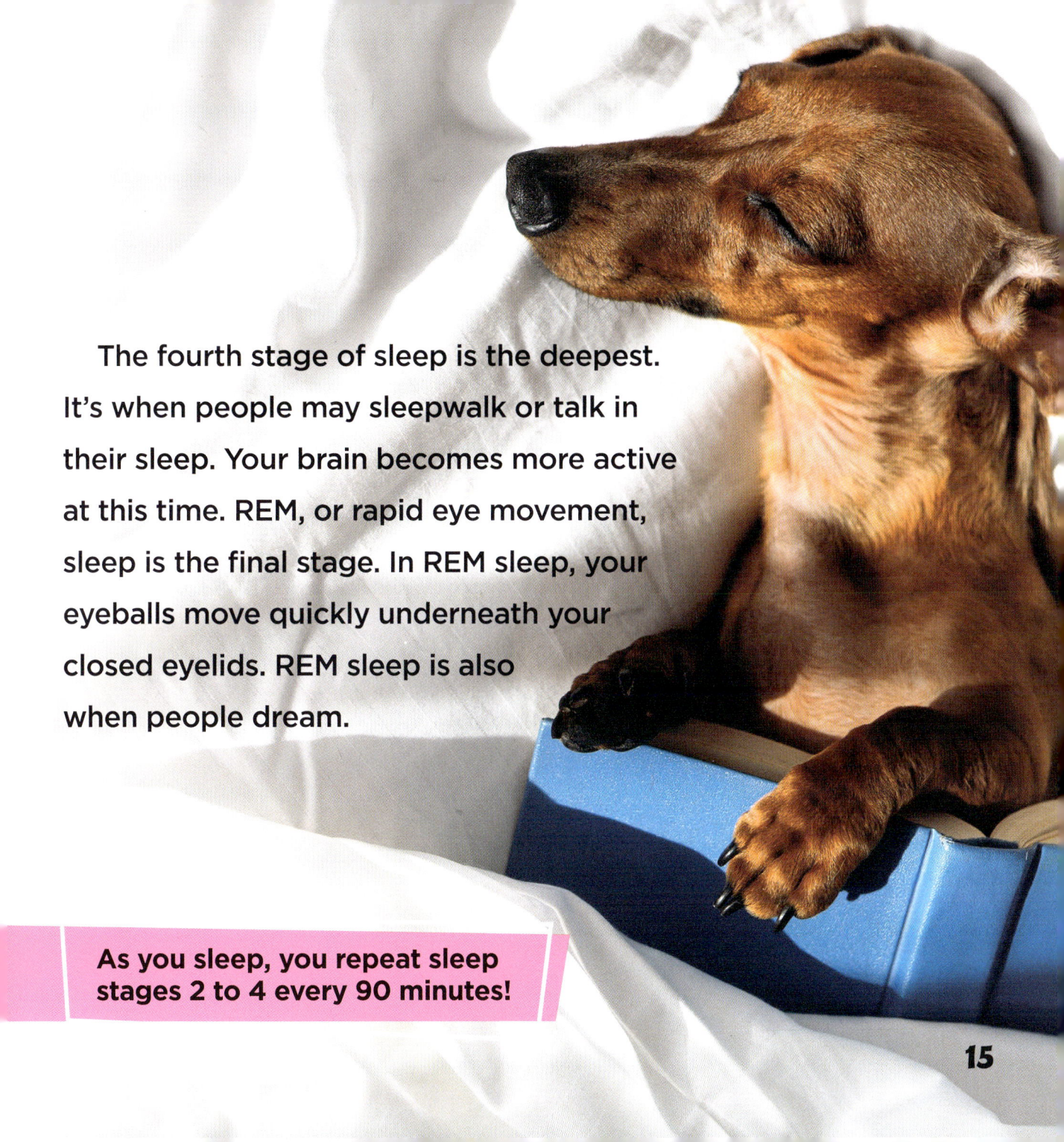

The fourth stage of sleep is the deepest. It's when people may sleepwalk or talk in their sleep. Your brain becomes more active at this time. REM, or rapid eye movement, sleep is the final stage. In REM sleep, your eyeballs move quickly underneath your closed eyelids. REM sleep is also when people dream.

> As you sleep, you repeat sleep stages 2 to 4 every 90 minutes!

SWEET DREAMS

A dream is a series of thoughts, images, or feelings. Dreams often star people you feel strongly about, such as your parents or siblings. Dreams can also be about something you're nervous or excited about, like the first day of school. Or they can be about something super silly, like a roller-skating unicorn or a monster doughnut!

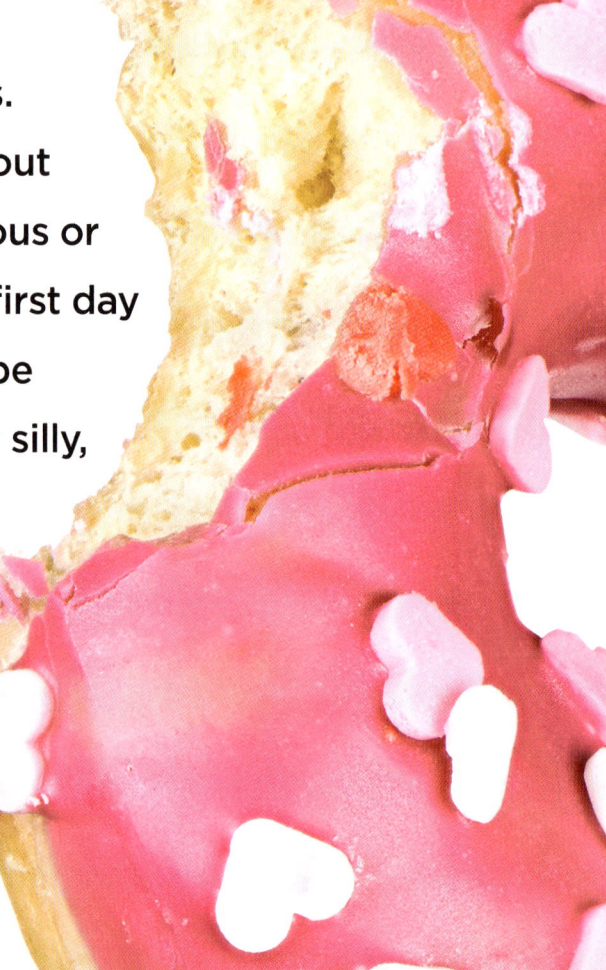

Let's say you had a dream about chasing a giant donut rolling down the street. Have you ever wondered why you didn't just get up from your bed and run after it? During REM sleep, your brain **paralyzes** your muscles. It's your brain's way of keeping you from hurting yourself during sleep.

A nightmare is a really intense or scary dream. The two most common nightmares involve falling and being chased.

Most people dream for a couple hours each night. But remembering those dreams can be a challenge. You're mostly likely to remember a dream if you wake up as you're dreaming it!

Why people dream is a big mystery to scientists. Some experts believe that dreams are a way for your brain to process your experiences from the day. Others, like the famous doctor Sigmund Freud, thought that dreams provided clues to how a person feels deep down.

There's no way of knowing what exactly dreams are for. However, experts do know how critical it is to get plenty of good-quality sleep. Going to bed at the same time every night can help, as well as turning off your TV and phone well before bedtime. Also, use your bed for sleeping. This helps train your brain that beds are only for sleeping—not for homework or reading.

Author Mary Shelley wrote her most famous book, *Frankenstein*, after an intense dream she had.

SLEEP ISSUES

Not everyone sleeps well. Millions of people suffer from insomnia (in-SOHM-nee-uh), the inability to fall and stay asleep. Insomnia has different causes, including stress and certain medicines. It can last for days or weeks but often goes away on its own.

On the flip side, people with narcolepsy (NAHR-kuh-lep-see) experience an overwhelming urge to sleep. In the worst cases, they can fall asleep while talking or walking. The good news is there are many treatments to help people improve their sleep—and care for their bodies and brains!

Melatonin and serotonin are chemicals made by the brain that help a person sleep.

BRAIN GAMES

For the next seven days, keep a sleep and dream journal! Every night and morning, answer the following questions in a notebook.

◆ At night, briefly describe your day. What were the highs and lows?

◆ In the morning, write about your sleep. Did you wake up during the night? Do you remember dreaming? What did you dream about?

At the end of the week, analyze your notes. Try to make connections between your experiences during the day, such as having a fight with a friend, and your sleep and dreams at night. What conclusions can you make?

GLOSSARY

chemical (KEH-muh-kuhl) a natural substance that helps the body function

demonic (dih-MAH-nik) relating to an evil spirit

disoriented (dis-AWR-ee-en-tuhd) confused as to a time or place

electrical (i-LEK-truh-kuhl) related to the flow of electricity, a form of energy

hypothalamus (hye-poh-THAH-luh-muhs) the part of the brain that's mainly involved in emotions

immune system (ih-MYOON SIH-stuhm) the system that a body uses to protect itself from diseases

organ (OR-guhn) a body part that does a particular job

paralyzes (pahr-uh-LYE-zuhz) the loss of movement in the body or a body part

replenishes (rih-PLEH-nish-uhz) makes full or complete again

synapses (SIH-nap-suhs) gaps between neurons where nerve impulses, or signals, travel

tissue (TIH-shoo) a group of similar cells that form a part of the body

vegetarian (veh-juh-TAIR-ee-uhn) a person who does not eat meat but may eat eggs and dairy

FIND OUT MORE

Books

Mason, Paul. *Your Mind-Bending Brain and Networking Nervous System*. New York, NY: Crabtree Publishing, 2016.

Silver, Donald M., and Patricia J. Wynne. *My First Book About the Brain*. Mineola, NY: Dover Publications, 2013.

Simon, Seymour. *Brain: Our Nervous System*. New York, NY: HarperCollins, 2006.

Websites

American Museum of Natural History: Brain Introduction
https://www.amnh.org/exhibitions/brain-the-inside-story/brain-introduction

The Franklin Institute: Your Brain
https://www.fi.edu/your-brain

KidsHealth: Nightmares
https://kidshealth.org/en/kids/nightmares.html

INDEX

brain stem, 13
cortex, 13
dreams, 5, 8, 15–19, 22
Foldvary-Schaefer, Nancy, 7
hypothalamus, 13
immune system, 12
insomnia, 20
muscles, 14, 16
narcolepsy, 21

neurons, 9
nightmares, 17
REM sleep, 15–16
sleep, 4–8, 10–15, 17, 19–22
sleepwalking, 4–7, 15
somnambulism, 6
stress and sleep, 20
synapses, 9
tissue, 13

ABOUT THE AUTHOR

Joyce Markovics has written hundreds of books for kids. She's fascinated by the human brain and all its complexities. She would like to dedicate this book to Anna, a dear friend and new mom, and baby Miles, who will one day sleep soundly through the night.